THE
PERFECT TIME

The Universal Calendar

Mark J. Silen

iUniverse, Inc.
Bloomington

The Perfect Time
The Universal Calendar

iUniverse books may be ordered through booksellers or by contacting:

iUniverse
1663 Liberty Drive
Bloomington, IN 47403
www.iuniverse.com
1-800-Authors (1-800-288-4677)

ISBN: 978-1-4502-7800-3 (pbk)
ISBN: 978-1-4502-7801-0 (ebk)

Library of Congress Control Number: 2010918107

Printed in the United States of America

iUniverse rev. date: 12/20/2010

Contents

Introduction

My name is Mark Silen. I have never written a book, and never in my wildest dreams did I ever think I would. But after encouragement from scientists, university professors, statisticians, mathematicians, authors, and others, I knew that I should write a book – about what I have come to call the perfect Universal Calendar. Why? Because, I have either made one of the most astounding discoveries about calendar miscalculations, or I have uncovered a coincidence beyond imagination.

Let me just put it bluntly – most of the world's calendars used today are flawed!

Calendars from farther back than any of us usually think about were instituted using incorrect astronomical calculations and what I call, 'ego-centered additions,' to the number of days per month causing them to be in error. I will go into how badly later.

There are many professionals who agree that the errors perpetuated, for many generations, not only should be corrected but would positively change the way we view several sciences. Why does this matter? The fact is that the flawed calendars we use daily have had adverse effects, both economically, and in science. As one Texas university mathematician said, "Mr. Silen's Universal Calendar may well be the answer to several scientific questions. It has the potential to change the way people do – just about everything; how products are marketed; how banks do their accounting, how scientists, and professionals, the world over, calculate and compute data from so many disciplines that it boggles the mind. The remarkable part of this whole thing is that Silen's Universal Calendar is so simple a child can use it."

Until I created my Universal Calendar, I really had no idea about all the different calendars, nor did I care, not unlike most of the people in this world. However, it appears that my Calendar is the most simple; logical and meaningful calendar ever developed. The big question now would be, how important is the truth about how calendars were instituted, and how would people react to it?

I presented my calendar to the public during different venues, for several years, and everyone that I encountered; including mathematicians and other professionals, who take these kinds of things more serious than I, convinced me that not only is my calendar more accurate and meaningful, but indeed ought to be adopted.

CHAPTER 1

'A COLD NOVEMBER NIGHT'

How I invented Silen's Universal Calendar

It was a very cold, icy, and wet November evening, in 1995, in Big Spring, Texas.

While being stuck indoors and nothing to watch on TV of any real interest, I started thinking about how one would create an imaginary world – for a computer game; a type of role playing game, if you will. I figured that the first thing this invented world might need was its very own time system; a calendar. So I began playing with numbers.

I tried a small variety of combinations, such as ten and twelve months of thirty days and a several other sets of numbers. After a little over an hour of playing around, I made an absolutely amazing discovery, yet so simple that it changed my life. I wondered if this discovery could change the lives of everyone else – the perfect calendar. Needless to say, the imaginary world I was going to create was never given a second thought afterwards. Never in a million years did I think that seven numbers; two easy addition equations, and six simple multiplication equations would apply to so many different disciplines.

Issues in religion, mathematics, science, history, nature, economics, and even calculations for monuments like the Great Pyramids in Egypt, or the Temple structures in Central and South America, could have been affected by the use of these seven, perfect numbers.

CHAPTER 2
'THE SON OF AN INVENTOR'

Being the son of an inventor, who sometimes wasn't so successful, caused me to protect myself first and foremost. I drew up the first concept of my Universal Calendar idea and had it copyrighted.

Even though I still wasn't taking this calendar very seriously, I thought to myself, I could still market it as a novelty. I didn't have to expect anyone to take it too seriously. It was just something to make you say, hmmm. But I have to admit, I found myself getting more excited about my little discovery than I was trying to avoid.

I grabbed a Dallas phone-book and happened to find a man who had a calendar printing and novelty store. He agreed to meet me right away. When I got there, he showed me the kinds of products he sold. He sold calendars, hats, shirts, mugs, pens, and pencils with company names on them. He basically had a small advertising company.

We went into his office so that I could show him my idea. At this point I hadn't done enough research. I just went in there half cocked and showed him the Universal Calendar I had in mind; the basic math, and a few of the applications that it could be used for. I also explained the novelty ideas I had in mind.

When I got finished, I was shocked by his reaction. He got so serious, and began talking like this was a real working calendar that needed to be adopted. He told me what my cost would be to have the Universal Calendar printed and to contact a marketing consultant.

He also asked me a very good question that I could not answer at the time. He asked me about birthdays of all things. How was I going to address people with birthdays beyond the 28th of the month? I was going to have to think about this one for a while.

I went straight to the phonebook and got in contact with a marketing consultant. It didn't take long to find a marketing consultant that would meet with me. The inventor spirit of my dad ran wild through my veins.

When it came time for my meeting, I put on an old, gray pen-striped suit, stuffed my briefcase with all my information; confidentiality agreements, and other pertinent documentation then proceeded to take my overzealous, but now slightly more aware self to this meeting. I got real nervous when I came to the address. This was one of those big sky scrapers, and the office was super nice. I am from Big Spring, a small west Texas town, and I began to feel like I was way out of my league.

As I sat in this fancy waiting room, I got more and more nervous. The secretary came and led me down this long hall to two huge wooden double doors. She seated me at large desk in front of a well manicured man who seemed friendly enough. He introduced himself as Sam. We shook hands, and I proceeded to tell him about my Universal Calendar.

When I was finished telling Sam about my ideas; marketing the calendar as a novelty, he frowned and seemed to not like the idea of the calendar as only a novelty.

That was my impression anyway.

Sam looked at me and began to laugh. "You can not even fathom what you have done here," he said. "If anyone has ever reinvented the wheel, you have come closer than anyone I have met," he added. Then said with a huge grin, "I'll bet it took you years to come up with this idea."

I said hesitantly, "No sir, it actually only took a few hours and truthfully, it was totally by accident." Sam jumped up and said loudly, "God gave you this calendar."

I calmly said to Sam, while inside I was totally freaking out at this point, "Nothing came out of the sky and said, "HERE," but if God put the thoughts in my mind, I won't deny it."

Sam then began to apologetically tell me his firm was not near big enough to take on such a global endeavor.

'Global endeavor?' I just wanted to market the Universal Calendar as a novelty. Sam would not even consider that idea. But now I was extremely excited as well as extremely confused.

The confusion came with thinking that I might have actually made some sort of a real discovery; and what if I did? But the Universal Calendar was just a novelty, or was it? I freaked out some more.

I went home and began some serious research. I had no idea I would find so many different subjects that would apply. I looked at the history, science, religion, nature, math, and so much more. In talking with friends and family about the Universal Calendar, it wasn't long before I found myself in front of a local news paper reporter; and later a local television station. Even the associated press picked it up. Everyone seemed to be taking the calendar too seriously.

After the story about the Universal Calendar appeared in the Big Spring Herald and in other newspapers, I met a local morning TV show host. Her name was Johnny Lou. She asked me if I would mind doing a two part segment on her show; I agreed.

Miss Johnny Lou was very supportive. She called me one day to come to a local political meeting, where she introduced me to local congressman, Charles Stenholm. After the meeting he met with me a few minutes, and I briefly explained my Universal Calendar.

He was very enthusiastic. He signed one of my calendars, which read: "Great idea, best wishes." – Charles Stenholm 118th district.

I was really beginning to think that I may have stumbled onto something big.

The response I got, from a wave of publicity was really quite phenomenal. I received calls from all kinds of people – from all over. Most of the response was very supportive and positive. However, there were a few that actually scared me. These people did not threaten me directly, but they would say things like, "There's 'nut-jobs' out there who think about shooting you, if you keep pushing your Calendar," one person said. I took some of these warnings seriously.

In my mind, my calendar was still only a novelty; I couldn't understand that kind of anger. Mostly though the response was positive; it heartened me.

My calendar has many applications, so I couldn't believe my discovery, using only seven numbers, was exclusive. And that if it was so simple, someone else surely had figured it out – right? Wrong. The more I researched, the more I found that very few had even attempted to right the wrongs in our calendar system, and every calendar that I researched was found to be lacking.

The Gregorian calendar system for instance, which is the system we use today in America, as well as in other countries, calculates the number of days in a year using flawed astronomical mathematics, not to mention the fact that at least two Roman emperors added days to the calendar. Let me explain:

There are 52 weeks in a year; seven days in a week. This equals 364, not 365 and a quarter. I counted the first day of winter to the first day of spring on the calendars we use today. There were ninety-one days. Four seasons, times 91 days, is 364; again, not 365 and a quarter. Then I read something about the "blue moon," – or the thirteenth moon. So I took thirteen months of 28 days, and this equaled the same 364 days, again, not 365 and a quarter, which is the numbers being used today – incorrectly. This really sparked my curiosity. I already knew that Leap Year was a day that men threw in to reconcile the 365 and a quarter days – which has been calculated as the Earth's orbit around the Sun.

Now I was raised to believe in GOD, and I don't profess to know more about who He is; but I do believe that GOD did not miss perfection by a day and a quarter. It must be that man was off by a day, or maybe someone else could have thrown in a day. I also thought, surely science has calculated these things; surely they must not be wrong. I became more curious the more research I did.

My journey began at a local library. As I studied, first how our present calendar system came to be then how other calendars from around the world were compiled, calculated, and used. I found that aside from several articles in various encyclopedias, there wasn't a single book about calendars; how they were instituted, who began calculating the dates and times; and how we ended up with the calendars we are presently using.

The Internet had a few web-sites on the subject, but they were so similar in substance – it left much to be desired, I gave up looking for awhile.

Some of my sources include the World Book and Compton's Encyclopedia; a web-site called, "Calendars and their History," which was produced by a man named L. E. Doggett, and his article: "Explanatory Supplement to the Astronomical Almanac," from the University Science Books, (Sausalito, California); and a 1996 Farmers Almanac, and other comparable information.

One of the first things that grabbed my attention was a statement which I read in Compton's: "The making of an exact calendar has perplexed mankind for ages because the division of time by days, weeks, months, and years, does not seem to fit together properly."

According to my calculations, as well as those by present-day astronomers, nearly two extra days were added to the calendar – mainly because of a simple error which has been assumed by – well, everyone; that a day is exactly twenty-four hours long; the fact is that a day is actually 23 hours and 56 minutes long, give or take a second or two. This is an average of four minutes per day which has been added to the calendar. After 52 weeks, 22 hours and 24 minutes has been added; nearly one more day. Then there is the quarter day to account for. So-called scientists, living in medieval times that also thought the earth was flat, added this quarter day because of flawed astronomical calculations.

There are two methods for calculating the length of a year: The time it takes for the Earth to revolve around the sun, and the time it takes the moon to revolve around the Earth. Some of the "miscalculations," resulted in a "Leap Year."

CHAPTER 3
THE HISTORY AND EVOLUTION OF CALENDARS

L.E. Doggett stated, "Calendars are for religious and social needs of a civilization or culture, and whatever their scientific sophistication, calendars must ultimately be judged as social and religious contracts and not as scientific treatises."

He went on to state that because calendars are created for social and religious needs, the question of a calendar's accuracy is "usually misleading and misguided." Could the accuracy of today's calendars really be misleading or misguided? Resoundingly, yes!

Statements such as the above by L.E. Doggett provided enough inspiration for me to continue researching. So far I have already seen so many discrepancies in the information I found on calendars that I had to continue.

Compton's states, that the Sumerians were the first civilization to create a calendar – among other notable items. This calendar was made up of twelve lunar months, consisting of 355 days long, and had a thirteenth month, every four years. The Egyptians, Greeks, and Semitic peoples, copied that calendar system.

Later, an Egyptian astronomer named Sosigenes came up with the first "solar calendar," – much like the one we use today. The early Romans also had a calendar, but it was based on the moon's cycle, which is 355 days around the earth; they also added a thirteenth month – every four years.

In 46 B.C., Julius Caesar adopted the 365 day calendar used by the Egyptians, created by Sosigenes. His calendar would be known as the Julian calendar and be in use for the next 1600 years. He also He renamed the month of Sextilis to July, because according to his 'scientist's observations,' it was the longest month of the year. Seventy years later, Augustus Caesar, (Julius Caesar's nephew) would rename the month of Quintiles to August.

I have read of a legend, (though I can not corroborate it) that states Julius Caesar not only renamed the month of July after himself, but may have thrown in an extra day – wanting it to be the longest month of the year. However, according to Compton's, July already had thirty-one days, and it was Augustus Caesar that threw in the extra day in August to match that of July.

I asked myself, could this possibly be the extra day that was added to calendars?

I found that in those days, Caesars' were considered to be human gods. If either Julius or Augustus wanted to throw in a day during any particular part of the calendar, they could, and probably did. The egos and vanity of these men, allows little doubt that the extra day in our calendar was due to their self-centered desires.

In 1582 A.D., or about 1600 years later, the Vernal Equinox, or first day of spring, happened to fall on March 11th instead of March 21st, the correct date.

Pope Gregory XIII ordered that 10 days be removed from the calendar, making October 4th, October 15th. He then ordered that one day be added to February every four years, except three times every four hundred years, thus creating the Leap Year. This new calendar would be called the Gregorian calendar, and is the very calendar we use today. We're not exactly sure why this pope made these changes. In my research I found that England did not adopt the Gregorian calendar until 1752, and that eleven days had to be dropped. The Eastern Orthodox Church did not adopt the Gregorian calendar until 1923, and in their case, thirteen days were "lost."

There were several other calendars which expressed the differences in calculating time: There was the Hebrew calendar, an Islamic calendar, and a Chinese calendar, as well as several European calendars. In the Americas, the Aztecs, Mayan, Inca, and others, each had their own forms of calendar systems. According to one Internet source, the Hebrew calendar determined that the world was created in 3761B.C. – that date of course, was calculated using the Gregorian calendar system. We think this source may have been a bit "touched."

Their Hebrew calendar is a lunar calendar which has twelve months, with a thirteenth month inserted seven times, every nineteen years. The procedure to figure out when this thirteenth month falls, according to several sources comes from ancient Babylonian tradition. Other adjustments periodically made to ensure that the Passover Feast, which follows the first day of spring, were also made.

The Islamic calendar is strictly a lunar calendar system, in accordance with the teachings of the Qur'an. This calendar consists of 354 or 355 days per year. The twelve months in this calendar alternate between twenty nine and thirty days. The Islamic calendar has no thirteenth month and the New Year can occur in any season.

The World Book Encyclopedia explains that the Islamic calendar divides time into thirty year cycles; with nineteen years having 354 days per year. Eleven months would have 355 days in a year; this method of counting days, weeks and months are about as accurate as the Gregorian calendar.

The Chinese did not adopt the Gregorian calendar until 1912 and no days were "lost." The Chinese however still use an ancient calendar, which is a lunar calendar, made up of 354 days. It has twelve months of alternating between twenty nine and thirty days, similar to the Islamic calendar, with a thirteenth month inserted like the Hebrew calendar.

The western cultures, such as the Inca, Aztec, and Mayan had entirely different calendar systems altogether. However, they served as useful tools just as the other calendars discussed above. The Aztec and Mayan calendars both consist of 365 days, made up of eighteen months and twenty days long, with five "evil omen days." However, the Aztec and Mayan years did not coincide. These years were also thought to have made a 52-year cycle. The Mayan calendar had 360 named days, and a year of 365 days. This formed a cycle which was longer; 18,980 days or 52 years of 365days, called a "calendar round."

The Inca calendar was the strangest one of all – to me. This calendar was made up of a nine day week, three weeks long, which is twenty-seven days. Twelve months normally made a year, but a thirteenth month was placed every third year. This formed twenty cycles of thirty-seven moons which formed a 60-year period.

Then I found a website on "calendar reform," which explained the use of a thirteen month calendar, from the late 1890s to the early1900s – called the Cotsworth calendar.

The 13th month was to be named "Sol," and was to be inserted between June and July. The365th day would be at the end of December and known as "Year Day." It would not have a week day name. "Leap Day" would be moved from February, to fall between June 28th and the new month of Sol. This was primarily presented as a business calendar. Other calendars have a variety of lengths, from an early Roman calendar of 304 days; to the Incan calendar of 324 and 351 day years then the more common length of 354, and 355, 365, and 365 and a quarter day – but no mention of a 364 day year.

Doggett explained that as of 1987 there were about 40 different calendars in use around the world. Furthermore, there have been over seventy calendars in use throughout history.

Christmas and New Years had come and gone. It was now 1996.

I was standing in line at a checkout counter at our local grocer, when the 1996 Farmers Almanac caught my attention. I began to flip through it when I saw an article that quickly added fuel to my research fire. The article was called Calendar Complications, "When is the Real First Day of Spring, March 20th or March 21st?"

This article in the Farmers Almanac stated that the first day of spring, or Vernal Equinox landed on March 21st only 36 times in the last 100 years. It also said that since 1980, the first day of spring has been on March 20th – and in 2010, spring will begin even earlier. The first day of spring will occur on March 19th. This made me wonder if "Leap Year" had fixed anything at all, if the first day of spring seems to be moving backwards – yet.

The article stated that summer had 93.641 days; fall had 89.834 days, winter had 88.994 days, and spring had 92.771 days in it. This made absolutely no sense to me. When I counted on the calendar we use, from the first day of winter to the first day of spring, there were ninety-one days. Ninety-one days times four seasons is three hundred sixty-four, not three hundred sixty-five. Which season is a day, or a day and a quarter longer?

Another area I want to address is the different types of years.

The lunar year is 354 days and sometimes, 355. Then there are three types of solar years, the sidereal year, the tropical year, and the anomalistic year.

The sidereal year is calculated when the Earth aligns with the Sun and a particular star. This year is measured between two successive crossings. The sidereal year is 365 days, five hours, nine minutes, and nine and a half seconds. The tropical year is calculated when the Earth's axis is at a right angle to a line from the sun. This happens twice a year on the Spring and Fall equinoxes. The tropical year is 365 days, five hours, forty-eight minutes, forty-six seconds.

Because the seasons are relevant to the position of the Earth's axis, the calendar we are using today is based on the tropical year. The anomalistic year is calculated from the point in the Earth's orbit which is closest to the sun. This type of year is the longest and is 365 days, six hours, thirteen minutes, and fifty-three seconds.

One thing this particular article failed to mention was the length of a day. As mentioned before, a day is not actually 24 hours, but 23 hours and 56 minutes long. Therefore, a day is four minutes short of 24 hours. If you account for this time every year, a day is lost.

In my research on calendars, I have learned many things, such as how many different calendars there are, with lengths of 354 days, or 355 days, or 360 days, to 365 days, and our very own 365 and a quarter day, calendar. I have also learned that the accuracy of calendars is not only 'very misleading and misguided,' the very accuracy of our calendar is questionable.

This was all very scientific to me – and like others, it was almost too much information to handle. I still wondered whether my Universal Calendar was the correct way to go or not.

Remember, according to L.E. Doggett, "Calendars are social contracts not scientific treatises." There were other facts, such as the article from the Farmer's Almanac, about the real first day of spring, and L.E. Doggett's statements about calendars being 'misleading and misguided.' There was the fact that all the different calendars had different length of days, months, and years, and as Compton's stated, "A perfect calendar does not seem to exist."

Until now!

CHAPTER 4
THE CALCULATIONS

This is how my Universal Calendar works:

There are 7 days in a week, 52 weeks in a year. This is 364 days, not 365 and a quarter.

(52 x 7 = 364) I then took a regular calendar and counted the first day of winter to the first day of spring, and there where 91 days. Well, 4 seasons of 91 days each is the same 364 days, not 365 and a quarter. (91 x 4 = 364) Now after playing with a little math I realized that 13 months made up of 28 days each was the same 364, not 365 and a quarter. (13 x 28 = 364)

Since Pope Gregory had thrown in Leap Year, and either Julius or Augustus Caesar probably did throw in an extra day, and after reading about the Cotsworth calendar, I stayed firm when I decided to throw that extra day and a quarter out.

This also would give Silen's Universal Calendar its uniqueness.

The next big decision that had to be made would be what to name the 13th month.

The Cotsworth calendar called its month "Sol," and placed it between June and July. This did not appeal to me whatsoever. The thirteen month should have just as much meaning as my calendar.

After giving this subject some serious thought, I finally found a name that would have the meaning that the 13th month needed. The name of this month would be called "Remember," and for three reasons.

First: as a religious tool this calendar has many religious applications which I will address in more depth "Remember" could be used as the first month of the year. Most religions around the world, such as Muslim, Jewish, and many Christian faiths believe that the true Sabbath day is in fact on Saturday. After all, Saturday "is" the seventh day of the week.

With the Universal Calendar every week, month, and year, ends on a Saturday. Everyone religious could "Remember the Sabbath."

Second: as a social tool, for the not so religious, Remember could be used as the last month; to remember the previous year, and to get ready for the next. I, being the more religious type, prefer that "Remember," fall on the first month, but I will leave it to the decision of those who might actually implement this calendar to decide.

Lastly: I decided to name the 13th month, Remember, because it simply rhymed with December. It had a kind of ring to it, November, December, and Remember.

Now there were two more questions that came to mind: What did this calendar really mean, and what to do with it?

If I could present a calendar that had so much more meaning than any other calendar in history, and show the simplicity of it, how willing would our society be to accept it? That was the question that has stayed with me ever since I discovered the flaws and the resolution for calculating an accurate formula to measure time.

CHAPTER 5
My Universal Calendar

Several years after setting the Universal Calendar aside, I was watching a TV documentary about the pyramids of Egypt. It was about the pyramids' history and how they were believed to be built. As I watched, I just kept wondering for some reason, could the pyramids have anything to do with my Universal Calendar?

I grabbed a pencil and some paper and began to play with the numbers. It took just a little while, and when I was finished the hair on my arms were standing straight up. It was then that I knew for sure that by sticking to my guns about not adding that extra day and a quarter to the Calendar, I knew without a shadow of doubt that I was on to something potentially huge. Now the inventors blood flow and his spirit, rushed back through my veins, and I was now ready to present my Universal Calendar to the world.

As I have mentioned, this Calendar consists of 364 days instead of the 365and a quarter days. There are 7 days in a week, and 52 weeks in a year, this equals 364 not 365 and a quarter. There are also 4 seasons, and when I counted from the first day of winter to the first day of spring on a regular calendar there where 91 days. Well, 4 seasons, times 91 days per season are the same 364, not 365 and a quarter days. When you take the months of the year and make each month 28 days each, you get an even 13 months. Well, 13 months times 28 days is the same 364 days not 365 and a quarter.

Here again is the argument over that extra day and quarter. First, we know for a fact that Pope Gregory XIII was responsible for adding a Leap Day. That leaves that one day that had to be accounted for. I have already explained that Julius Caesar or his nephew Augustus Caesar possibly and probably threw in a day, out of vanity. Then there is the science of that time with some stating that the Earth was flat and that the Earth was the center of the universe. We now know that these scientific statements of that time were not true, so could the science of that day have also been a day off?

As I have stated, I am a believer in GOD. And I asked, did GOD miss perfection by one day, or was this something mankind did? There are many religious applications which I will discuss later on in this book.

The Female Reproductive Cycle

Having been a licensed nurse for many years, I know that the complex female reproductive cycle also can be applied to the Universal Calendar. The normal menstrual cycle is a 28 day cycle, and a full term pregnancy is 40 weeks. Most women know they are actually pregnant 10 months not 9, and due dates would be so much easier to figure out too. From the day of conception to the due date would be a perfect 10 months.

Sliding scales would no longer be needed. If women alone accepted this calendar it would get adopted.

Birthdays

Silen's Universal Calendar may also give even more meaning to birthdays. The days would be fixed; whatever day one is born on, that day would always be the same, thus your true birth day. For example, if you were born on the 12th day of the month, your birthday would always be on a Thursday. It has already been mentioned about the concern over birthdays that now fall beyond the 28th day of the month if the Calendar was adopted. I have given this a lot of thought.

To be honest, and blunt, this is simply a temporary generational problem. After maybe 40 years or so this problem simply fades away. I thought of only one consolation to those born beyond the 28th day of the month. Like Leap Year babies, if you choose to celebrate your birthday on the 28th of the month, it will always be on a Saturday.

Banking and Book-keeping

I approached a bank officer at a reputable and local bank here in Big Spring – about obtaining a small business loan to get my Universal Calendar off the ground. After explaining the idea, the loan officer was amazed. She stated that this Universal Calendar could "fix all the problems associated with banking and book-keeping."

She wrote the following letter as a testimonial about her opinions:

"From a banker's standpoint, our present calendar has always been a nightmare. Do we use 365 days, 364 days, 360 days…or good- ole leap year = 366 days?

- **Maturity dates for loans and certificates**: There is confusion with maturity dates for loans and certificates of deposit: 90 days equals a quarter year. 180 days equals a half of a year. Both of these represent only 360 days total. 182 days is also a half of a year, which equates to 364 days for the year; what happened to 365 and a quarter days?
- **12 months works**, most of the time, but then you have Leap Year, or 366 days, there is always a problem.
- **This ongoing problem** with interest accrued for checking and savings accounts, certificates of deposit, as well as loans, takes special programming of our computer software. Our calendar also effects customer trust…they are not quite sure we paid them 366 days interest for their 12 month certificate of deposit for Leap Year and they want us to prove it to them.
- **If one receives** interest monthly from their certificate of deposit, it is a different amount each month. (It's figured on the number of days between pay periods…January 27th to February 27th, versus February 27th, to March 27th, will be a different number of

days.) The daily interest factor is arrived by dividing the annual interest by 365 days and 366 days on Leap Years. This is then multiplied by the number of days between pay periods.

- **Payroll problems:** "The interest check for January will be more than the interest rate for February, etc. So, at the maturity date, they say prove to me that I received interest for 365 days. Getting paid every other Wednesday, or every two weeks, creates more pay periods in a year than getting paid on the first and fifteenth of each month – which creates payroll problems.

Business

I then found an Internet site where George Eastman of the Eastman Kodak Company, also supported a thirteen -month calendar in 1926. His presentation was focused primarily on business rather than banking and book-keeping. He gave eight reasons for supporting a thirteen month calendar. At that time it was the business calendar of Cotsworth that he supported. However, his reasons could also apply to the Universal Calendar. Here are the eight reasons George Eastman supported the 13-month, Cotsworth calendar:

*All months would have the same number of days (28), the same number of working days, except holidays, and the same number of Sundays.

*All months would have four weeks.

*Each week- day would always occur on the same four fixed dates of the month.

* Quarter-years and half-years would be of the same length.

* The month would always end on Saturdays.

*A holiday would always occur on the same day of the week.

* The date of Easter would become fixed.

* Yearly calendars would no longer be necessary, one fixed monthly calendar would be sufficient.

I suspect if George Eastman were alive today, he'd support my Universal Calendar.

Ecological Reasons

The Universal Calendar could also have a major impact on the ecology of the Earth.

Think about how many sheets of paper it takes to produce one calendar today. Each calendar uses a minimum of twelve sheets of paper. Then think about all of the calendars in use around the world. That is an extremely large amount of paper.

The Universal Calendar uses only one sheet of paper, or better yet one piece of plastic. Since the Universal Calendar never changes, one calendar could be used year after year.

The Universal Calendar was designed to be used with an erasable marker. Just write in the year, circle the month and day, and the Universal Calendar will last time and time again. Think about how much paper that could actually be saved in just a short time; and how many trees could we save just by switching to this calendar. This may even fall under the Paper Reduction act.

Simple Mathematical Systems

My Universal Calendar uses a very simple mathematical system consisting of only seven numbers, (4, 7, 13, 28, 52, 91, and 364) that also shows up in a wide range of gaming pastimes, and other such as ordinary playing cards, music, numerology, and the mystical pyramids.

A common deck of playing cards has the exact same mathematic principles as the Universal Calendar. There are 52 cards in standard deck with 4 suits of 13 cards each. Multiply each of these numbers and all the numbers come out in the same equation principles.

52 x 7 = 364, 4 x 7 = 28, and 13 x 7 = 91.

Many magic tricks are preformed using a regular deck of cards. There is one card demonstration in particular that is simple and works every time. I'll share it with you, because there really is no trick – it's simply a mathematical principle:

Take a regular deck of playing cards then arrange them in order and by suit. Now, cut the deck in half anywhere placing the top half over the bottom. Do this thirteen times. Then deal thirteen cards face down; finish out the deck so that you have thirteen piles of four cards each. Have someone choose a pile then turn it over. If done correctly it will be four of a kind. Flip the rest of them over and they will all be four of a kind and in numerical order.

The exact history of cards is unknown, but it believed to have originated in Egypt. It is also believed that the tarot cards used in fortune telling evolved from the more simple deck of playing cards.

Other games like dominoes and dice apply. A set of dominoes made up of 28 tiles using 7 numbers, played in 4 directions. With dice the math is slightly different. Even though there are seven numbers making up the Universal Calendar, there are actually only six equations at use in this system. So even though a die only has six sides, each opposite side equals seven. (1 and 6, 2 and 5, and 3 and 4) I know this is all purely coincidental, but it was fun when I thought about it.

The musical scale could also apply to the Universal Calendar. A musical scale is made up of eight whole notes and five half notes which is a total of thirteen notes…C, C sharp – D flat, D, 1, 2, 3, D sharp – E flat, E, F, F sharp – G flat, G, G sharp – 4, 5, 6, 7, 8, 9, A flat, A, A sharp, B flat, B, and C. 10, 11, 12, 13. At this point, I realized one more aspect of all this and how it might relate to the Universal Calendar and the thirteenth month of Remember being either the first or the last month. In a musical scale the 'C' note is the first note of the scale as well as the last note, however the last note of one scale is the beginning of the next, and also in a deck of cards. I know. It's just another coincidence.

The Pyramids

The pyramids in Egypt and elsewhere have been shrouded in mystery and intrigue for thousands of years. They have captivated the attention and imaginations of millions of people from around the world. From scholar to lay-person, people love to speculate who, how, when, and even why were the pyramids built. I do not claim to have the answers to any of these questions. In fact I will probably add more questions regarding the mysteries of the pyramids. However, you may see how uncannily they relate to my calendar.

There are pyramids built all around the world, in Egypt, China, and Central and South America. In this book I primarily address the pyramids of Egypt and the America's.

I mentioned earlier that I had been watching a documentary on television about the history of the pyramids of Egypt. For some reason, I wondered about a possible connection to the pyramids and the Universal Calendar. That is when I grabbed my pencil and a piece of paper and did some simple math, and I do mean simple. When I finished the hair was standing straight up on my arms, and I knew I was onto something possibly huge now.

I took the numbers 1 thru 7 and added them together, (1 + 2 + 3 + 4 + 5 + 6 + 7) this equaled 28. A pyramid has four sides and 7 x 4 is also 28, the same amount of days in a Universal Calendar month; coincidence? Then I took the numbers, 1 thru 13 and added them together. (1+2+3+4+5+6+7+8+9+10+11+12+13) This equaled 91, the same amount of days in a season with the Universal Calendar.

Again the four sides of the pyramid came into mind. 13 x 4 equals 52, the same number of weeks in a year, and 91 x 4 = 364, the same amount of days in a Universal Calendar year. The exact same numbers that make up the Universal Calendar, also make up a perfect pyramid, and a common deck of playing cards. What are the odds?

I don't know much about the "all-seeing-eye of the pyramid," I do know a little about it and that it is depicted on our very own dollar bill; is there something to that? My point is that if you look at the diagram I have laid out below, the numbers 1 thru 7, depict the top half or the "all seeing eye" and the numbers 8 thru 13, represents the base of the pyramid.

O
O O
O O O
O O O O
O O O O O
O O O O O O
7 O O O O O O O _28
O O O O O O O O
O O O O O O O O O
O O O O O O O O O O
O O O O O O O O O O O
O O O O O O O O O O O O
13 O O O O O O O O O O O O O 91

The Great Pyramid of Giza in Egypt was reportedly built over 4,000 years ago, according to several mainstream sources. The angle of this pyramid is 51.6 degrees; less than half a degree off a perfect 52 degrees. Could settling and / or erosion have caused this difference? The Great Pyramid also happens to sit on 13 acres of land.

The Mayan pyramid in Mexico called El Castillo caught my attention. This particular pyramid has 91 steps on its four sides. They say it has one more step at the top, but could it be just a platform for the temple that sits atop of this pyramid. Twice a year during the spring equinox and the fall equinox, the sun moves across this pyramid forming the shadow of a snake crawling up this pyramid.

Thousands of people each year flock to this place to witness these events.

One major question comes to my mind when I think of the math relating to the pyramids. Could this simple math have been the inspiration and calculations for building the pyramids?

Now I'm wondering what are the odds of seven numbers 4, 7, 13, 28, 52, 91, and 364; six simple multiplication, and two addition equations, $7 \times 4 = 28$, $13 \times 4 = 52$, $7 \times 13 = 91$, $4 \times 91 = 364$, $7 \times 52 = 364$, $13 \times 28 = 364$ — $1 + 2 + 3 + 4 + 5 + 6 + 7 = 28$ — $1+2+3+4+5+6+7+8+9 +10 +11+12 +13 = 91$ — creating a perfect calendar, a perfect pyramid, a common deck of cards, and applying to all the things I have explained and all the things I have left to explain.

The Universal Calendar also applies to areas of entertainment that captivate the imagination of millions of people around the world, and that is numerology and superstitions. I am not a numerologist but the numbers I am about to discuss, everyone is familiar with them, the numbers seven and thirteen. First, let us address that good ole number, "lucky number seven."

The number seven has been said to be a holy number signifying "divine completion," and is often mentioned throughout the Bible. The number seven is used significantly in many other ways. For example... the seven wonders of the world. Sail the seven seas; the seven continents of the world, the constellation, and their seven sisters. In some cultures we are the seventh planet in the heavens if you count from Pluto to Earth; seven colors in a rainbow, or being "the seventh son or daughter."

My point is that seven is a number that almost everyone will agree, is surrounded by a type of mysticism – It is a very significant number in numerology, and the Universal Calendar just happens to be made up of seven numbers.

Another number which has been an enigma for millions of people is that dreaded number "unlucky number thirteen." This one number has affected more people than any other number in the infinite supply of numbers. Even tall buildings do not have a thirteenth floor. Most people do believe the number thirteen is a very unlucky number, but there are a few of those who actually believe that thirteen is one of those lucky numbers. Many good things have the number thirteen. Jesus and twelve disciples are actually thirteen individuals who founded the religion of Christianity.

The great nation of this United States was founded by thirteen original colonies, and we have thirteen stripes on the flag in remembrance of this. My grand-mother once told me that a dozen used to be thirteen but twelve made packaging easier. Today we call this a baker's dozen. The "thirteenth." This day alone affects millions of people each year in various ways.

My Universal Calendar may help to clear this phenomenon up.

With the Universal Calendar every second Friday of every month is always Friday the Thirteenth. This is a natural occurring event that should be there every month, so when it occurs on our current calendars, the natural receptors of some people go off, and they don't know why; they call it superstition.

Geometry

I also noticed in my dabbling that with this calendar that every basic shape in geometry can be illustrated in the calendar.

The square, represented by the four thirteen week or ninety-one day seasons.

The rectangle, represented by four seven day weeks to form a month.

The circle, represented by the thirteen months revolving around the sun to form a year.

The triangle, represented by adding the number of days in a week, and the number of months in a year.

The pyramid, is represented when you multiply the triangle by the four seasons.

I had decided to take all of this information to my small town local college, Howard County Community College – just to see what a math instructor might think about all of this. I also wanted to find out if they thought all of this was some huge discovery or just one colossal coincidence.

I met with a Mathematics Instructor by the name of Dr. Linda Buchanan. I showed her all of the information I had gathered and all of its many applications. Her response was nothing short of amazement. She told me that she could tell right away that "the entire math was very simple and correct."

Concerning the issue of whether the Universal Calendar was a coincidence or discovery, she explained she was one of those people who did not believe in coincidences; therefore she thought that the Universal Calendar was probably more of a major discovery. Dr. Buchanan was then not very sure as to where I should take the Universal Calendar next. After giving it some thought for a while, she strongly suggested that I take this information to a theologian because the religious implications were really quite phenomenal.

CHAPTER 6
RELIGIOUS APPLICATIONS

There are several reasons why I have not discussed the many religious applications relating to the Universal Calendar.

First, because there are so many religious applications, I thought many might believe I had ulterior motives about pushing a particular religious belief, I do not. However, I wanted to keep the religious applications together, and in my opinion, to save the best for last; to keep the response of ministers and other religious subjects together.

Being raised in a Christian environment, I discovered a few of the religious applications to the Universal Calendar myself. The Universal Calendar applies to many denominations and possibly nearly all religions as well.

I have already mentioned that Saturday is always the seventh day of the week with the Universal Calendar. Every week, month, and year would end on a Saturday. Saturday would become the very hub of the calendar revolving around the "Sabbath day."

Genesis 1:14 reads:

"And God said, Let there be lights in the firmament of the heavens to separate the day and night; and let them be for signs and for seasons and for days and for years."

The first rule of any religion is, or ought to be is that God is perfect. Did God miss creating the 'Perfect Time' by just one day? I personally do not think so.

In II Peter 3:8 it reads: "But do not forget this one thing, dear friends: With the Lord a day is as a thousand years, and a thousand years is as a day."

Remember it was Pope Gregory XIII who threw in Leap Year. Men threw in other days and took some away – was the science of that day off? Obviously; some so-called scientists of those days proclaimed that the earth was flat and that it was also the center of the universe; that was only 500 years ago.

Now, if we take that one day out that was possibly thrown in out of vanity, we just happen to get a perfect calendar that would technically last a thousand years without changing, which would back up religious beliefs and faiths. Is it just another coincidence that the only holiday that falls beyond the 28th day of the month, is Halloween? This one day referred to by many as, "The Devil's Day," just happens to fall off.

The Universal Calendar does not take a single thing from religions; the Universal Calendar strengthens them.

I gathered up all of the information I had, put it together and took it as a rough package to a minister in here in Big Spring, Texas; Dr. Jimmy Watson. I met Dr. Watson through the president of Howard County Community College. Immediately Dr. Watson was very enthusiastic and supportive. He agreed that the best way he could help me would be to write a paper on the subject and allow me to publish many of his opinions. The following is his treatise and opinions:

Silen's Universal Calendar

"For the past several years Mark J. Silen of Big Spring, Texas has been promoting the idea of a new calendar consisting of 364 days in a year. One reason this calendar contains only 364 days and not 365 and a quarter day is based on the fact that many days have been added or subtracted to from our current calendar over the years. This calendar which has been dubbed, "Silen's Universal Calendar," is broken down into thirteen months, each consisting of 28days.

Calendars are based on three different types of years. The Gregorian calendar, for example, is based on a solar year, which is the time it takes the earth to travel around the sun; The Islamic calendar is based on a lunar cycle, bearing no relation whatsoever to a solar year. The Hebrew and Chinese calendars are based on a Luna-solar year. The years are based on the Earth's rotation around the sun, but the months are based on lunar months. How one measures a solar year from a scientific standpoint also varies.

Attempts at Calendar Reform:

Perhaps the human race is doomed to mere estimations concerning the calendar. However, Mr. Silen has noted in his research; "The making of exact calendars has perplexed mankind for ages – because of the divisions of time by weeks, months, days, and even years; none seem to fit together properly. Calendar reform, however, would be an extraordinary accomplishment even though, for example, the legal code of the United States does not specify an official nation calendar.

There have been attempts at calendar reform:. In the 1890's the "Cotsworth" calendar was developed, based on thirteen months. A month called "Sol" was to be inserted between June and July. The extra day of Leap Year was not to be removed. The argument can be made that new calendar reform is necessary at the very least for the purpose of simplifying business practices around the world including gathering ambiguous statistics for production and sales and creating constant pay periods. A "universal" calendar (a calendar for all societies) would simplify everything from school years to religious and government holidays to business accounting. Unfortunately, people are naturally resistant to calendar reform, probably because they perceive changes in traditional economic, religious, and social activities.

"Together, the arguments for a 364 day calendar from a historical-scientific and faith perspective go far in validating Silen's Universal Calendar." – Dr. Jimmy Watson

As noted in the introduction, the Universal Calendar created by Mark Silen, is a thirteen-month calendar based on twenty-eight days per month rather than twelve month calendar based on months of varying length. The logic behind this calendar is very simple. Every month begins on the same day of the week, Sunday, and ends on a Saturday.

There are still fifty-two weeks in a year and seven days in a week. Obviously, fifty-two multiplied by seven days is 364 days. Divided by the four seasons there are ninety-one days per season. This means there are thirteen weeks in a season or quarter.

In all there are only seven numbers used in this simple mathematical scheme: four (seasons), seven (days per week), thirteen (months per year; weeks per season or quarter), twenty-eight (days per month), fifty-two (weeks per year), ninety-one (days per season), and 364 (days per year). The Universal Calendar is one and a quarter day shorter than the current Gregorian calendar.

"Silen offers two reasons for maintaining a 364 day calendar. First, from a historical-scientific perspective, even though the various methods of measuring a year are in relative agreement that a year is roughly 365 and a quarter days, history has borne out the fact that this will average as being too long. The fact that many days at a time have been deleted from the Gregorian calendar in order to "fix" the "shifting" vernal equinox bears witness to this problem."

I might add the 1996 Farmers Almanac contains an article that states the first day of spring used to be on the 21st, at the time of this writing, it was the 20th, and in 2010 the first day of spring will be on the 19th. Does Leap Year really help?

"From a religious perspective, Silen's Universal Calendar is in itself a faith statement concerning the order, harmony, and stability of the created cosmos. Together, the arguments for a 364 day calendar from a historical-scientific and faith perspective go far in validating the Universal Calendar."

Pyramids and the Universal Calendar:

As discussed above, there are only seven numbers utilized by Silen's Universal Calendar. That Silen discovered a mathematical connection between these numbers and the construction of a four-sided pyramid as seen in the chart earlier illustrated in this book, the math is intriguing.

If the middle-row contains seven circles or dots per side, this equals twenty-eight circles around the middle of the pyramid. This correlates with the number of days in a week (7) in the Universal Calendar and the number of days in a month (28).

Mathematical Principles Inherent in the Universal Calendar

It also equals twenty-eight circles or dots per side in the top half of the pyramid. The equation looks like this: 1+2+3+4+5+6+7 = 28.

At the base of that same pyramid will be a row of thirteen circles or dots per side. Multiplied by four sides this equals fifty-two circles or dots around the base. Again the correlation is obvious. In the Universal Calendar there are four seasons of thirteen weeks each. This equals fifty-two weeks per year.

Also the face of each side of the pyramid contains ninety-one circles or dots, equal to the number of days per season in the Universal Calendar. We can now continue with the equation above: 28+8+9+10+11+12+13 = 91. All four sides of the pyramid added together (91x 4) equates to 364 circles or dots covering the outer surface of the pyramid. This equals the total number of days in a year in the Universal Calendar.

There are pyramids all around the world: in Egypt, South America, Mexico, and China. The Great Pyramid is a perfect pyramid. The angle of the slope is almost exactly a perfect fifty-two degrees. At El Castillo, a pyramid in Mexico, the sun climbs up the edge of the steps to form a snake climbing up the side every fall and spring. Thousands of people gather there each year to witness this event. Is it a coincidence that there are ninety-one steps on four sides, equaling a total of 364 steps.

There is one step up to a platform or altar. This is just a platform or alter in which sacrifices were performed.

Is it possible that the math depicted in Silen's chart is the mathematical scheme used to construct the ancient pyramids? Could this simple formula be the common denominator linking the building of all the pyramids? The math is simple enough for even ancient societies to grasp. The purpose of the pyramids was to entomb kings and emperors as in Egypt or temples as in Mexico. The shape is said to represent the rays of the sun. However, perhaps these pyramids serve as monuments to the very nature of time, including the possible divine origin of time.

Some scholars believe this Bible verse refers to the Great Pyramid at Giza. Isaiah19:19 In that day there will be an alter to the Lord in the midst of the land of Egypt, and a pillar to the Lord at its border.

Another mathematical coincidence related to the Universal Calendar is the mathematical scheme of a common deck of cards. Whether used for playing games or telling fortunes, the mathematical properties are the same. The relationship between a deck of cards and the Universal Calendar is simple: There are fifty-two weeks cards in a deck; there are fifty-two weeks in a year. There are thirteen cards per four suits in a deck; there is thirteen week in each of the four seasons in a year. Each of these numbers multiplied by seven (the number of days in a week) equals the other numbers related to the Universal Calendar: twenty-eight, ninety-one, and 364.

Other Numerological Phenomena and the Universal Calendar

At first glance the numbers used by the pyramids, a deck of cards, and the Universal Calendar seem inconsistent with traditional numerology. For Example, the number "3" is not utilized, nor is the number "12." However the number "7" is of primary importance in these formulas.

In Hebrew numerology, it has been known as the number of divine completion, or perfection, in contrast to the number "6," which is considered the number of denoting imperfection. Only seven numbers make up the entire Universal Calendar system.

The number "13" is also a surprise in the above-mentioned formulas. "13" is often considered to be an unlucky number. We avoid using the number "13" when labeling floor levels of tall buildings. We allow ourselves to become overly cautious on " Friday the 13th " when it occurs only a few times a year. However, in the Universal Calendar every second Friday is the thirteenth day of every month. Friday the 13th is supposed to be, and would be a naturally occurring event every month. This alone may influence people to disregard the Universal Calendar. However, in response to that possibility, Silen asks, "could our avoidance of the number '13' be like a subconscious guilt complex or something?" In other words, do we feel an inherent guilt for our avoidance as far as to create superstitions? He cites other examples of the avoidance of the number 13.

Thirteen was the original number of a dozen, now known as a "bakers dozen." Jesus and his twelve disciples; making thirteen individuals founding Christianity. There were thirteen original colonies founding the United States of America and it had thirteen stripes on the American flag to symbolize this. (Silen notes that if two more territories such as Puerto Rico and the Virgin Islands were to be granted statehood, the United States would contain fifty-two stars and thirteen stripes.) While much of this is probably coincidental, it does portray the significance of the mathematical scheme of the Universal Calendar already at work in our culture.

In his book, A Beginner's Guide to Constructing the Universe: The Mathematical Archetypes of Nature, Art, and Science, Michael Schneider discusses the relation of the number "12" to the number "13," and explains that any geometric construction, for example, of a decagon, or twelve- sided figure, will result in twelve points around a center point, or thirteenth point. Also, he notes that twelve spheres, packed together, will naturally create a thirteenth sphere in the center. Schneider's point can carefully be carried one step further, it would seem, to respond to the popular religious sentiment that the number twelve is sacred. For example, there were twelve tribes of Israel, and Jesus had twelve disciples. However, in the case of the twelve tribes, we could argue that God is the center point around which the twelve tribes revolve. The same argument could be made about Jesus and his relation to the twelve disciples. In this view, the number "13" becomes more appropriate because it represents the relationship between God and Israel, and Jesus and his disciples.

Implications of the Universal Calendar

Calendars are usually developed by people to serve one of two basic purposes: religious, or social Generally speaking, the purpose of calendar making is to organize units of time to satisfy these particular needs, they may not reflect exact science, but this is not their purpose. As L.E. Doggett suggests, "Calendars must ultimately be judged as social contracts, not as scientific treatises." In that sense, the Universal Calendar would be more efficient than the inconsistent calendars in use today.

There are several logistical problems that would have to be overcome in order to implement a thirteen month, twenty-eight day calendar. First, there is a problem with special days (holidays and birthdays) that fall after the 28th of any particular month. As far as holidays are concerned there is only one major holiday that occurs after the 28th: Halloween.

Some believe that it is either appropriate or coincidental that Halloween is fast becoming less and less popular, especially among members of the Christian faith. "2 Peter 3: 8 that with the Lord, one day is as a thousand years, and a thousand years is as a day." Is it not odd the one "Evil Day" falls off and you get a calendar that lasts a thousand years?" Whether this is warranted or not, the "fall festivals" that are quickly taking the place of traditional Halloween carnivals around the country would eventually make this problem moot.

On the other hand, Silen's Universal Calendar would make most other holiday observances more uniform and easy to remember. Christmas would always be on a Wednesday, Thanksgiving would always fall on the 26th of November, and Valentine's Day would always be on a Saturday – a perfect day for a romantic weekend! Easter would still vary because it always falls on the first Sunday following the full moon that falls on or after the vernal equinox, a date that would change from year to year, even according to the Universal Calendar.

As far as birthdays and anniversaries that occur after the 28th are concerned, this would only be a problem for one generation. After the Universal Calendar is implemented no one would be born with a birthday after the 28th and no anniversaries would occur after this date. In the meantime, folks with late month birthdays and anniversaries could celebrate them like Leap Year babies, on the 28th or the 1st. Silen suggests as a consolation, if one celebrates their birthday or anniversary on the 28th, it will always be on a Saturday. On a positive note, in the future birthdays and anniversaries would always fall on the same day of the week.

If people get married on a Saturday, then effectively choosing the day of the week they will always celebrate their anniversary. If a person is born on a Tuesday, then Tuesday would carry a special significance for them. It would be their "birth day." Certainly, we would be able to remember other people's birthdays and anniversaries more accurately if we were also familiar with a particular day of the week on which they fell.

There is also the problem of naming and placement of the thirteenth month. Silen proposes that we place the thirteenth month between December and January, calling it "Remember." He offers three main reasons for this name: 1) it could be used as the first month " to remember the Sabbath to keep it holy." Silen points out every week, month, and year would end on a Saturday making it the very hub of the calendar. 2) It could be used as the last month to remember the previous year. 3) 'It just had a "ring" to it,' he said: 'November, December, Remember.'

Those are all obstacles to overcome. On the other hand, the Universal Calendar provides opportunity to overcome other problems. For example; The Universal Calendar would coincide more naturally with the female reproductive cycle. The twenty-eight menstruation cycle could be more easily followed because there are exactly twenty-eight days every month in his calendar.

If the Universal Calendar was utilized the benefits to the environment would be enormous. Think of all those trees this calendar would save. Except for calendars to mark one's personal schedule, which now days is mainly electronic anyway. One sheet of paper or even plastic could be used over and over again. In fact we would all eventually memorize the calendar, since every date falls on the same day every month. An average calendar uses fourteen sheets of paper: a cover sheet, twelve monthly sheets, and a back cover. Millions of calendars are produced each year. How much paper could we save in one year alone if we adopted the Universal Calendar?

Another practical issue does concern the Sabbath.

Genesis 2: 3 reads, "So God blessed the seventh day and hallowed it, because on it God rested from all the work that he had done in creation."

In the Universal Calendar, every seventh day of the month is a Saturday. Silen states, " Every week, month, and year would end on a Saturday, meaning, Saturday would become the very hub of the calendar." This fits religious sentiments of many people around the world (even those who worship on Sunday) who are attracted to the idea of a "day of rest."

Silen's Universal Calendar, in all practicality, does not favor any one religion over another. In fact, Silen believes the Universal Calendar could strengthen all religions. Almost all religions around the world could fashion their religious holidays and traditions in accordance with the Universal Calendar.

The Roman Catholic and protestant churches already regulate their liturgical calendars according to the Gregorian calendar. A shift to the Universal Calendar would create much hardship. Perhaps the most resistance would come from the Islamic community, whose current calendar is strictly lunar. The Islamic year has twelve months with, alternately, 29 and 30 days, making a year 354 or 355 days. There is no attempt to align the lunar year with the solar year. Because of that, Muslim months and major festival observations have no relation to the seasons. Actually, this could make their adaptation of the Universal Calendar more closely coincide with the time period of the lunar months.

As noted earlier, the business community would benefit enormously from the Universal Calendar. Bookkeepers and accountants would have a much easier workload. Pay schedules would be uniform. Production and consumer figures could be more easily compared with previous months. Quarterly bookkeeping would be more precise. People who have to deal with the accounting problems of business, governmental agencies, and even family budgets, would find the Universal Calendar.

George Eastman's (Kodak) dream of a calendar reform for business purposes would be fulfilled. Finally, the confusion over how many days are in each month would come to an end. Not only is there confusion about the variety of calendars in history, there is also confusion concerning the fact that different months in our calendar contain a different number of days. As children we learn poems to help us memorize, for example, the fact that February contains twenty-eight days (except Leap Year), November uses thirty days and December has thirty-one. No more having to remember nursery rhymes or counting knuckles and spaces to figure out the months.

The Universal Calendar would bring convenience and consistency to many aspects of daily life.

Theological Implications

When we pause to consider the possible theological implications of the Universal Calendar we realize that we are simply grasping at possibilities. But the possibilities are intriguing. For example Isaiah 19:19 of the Hebrew scriptures speaks of "an alter to the Lord in the center of the land of Egypt, and a pillar to the Lord at its border."

The pyramids of Egypt had already been constructed before the time of Isaiah. As Isaiah writes in verse 20, could the pyramids, "be a sign and a witness to the Lord of hosts in the land of Egypt"? This presumes that the mathematical precision is somehow "divine" in origin and nature. This brings up philosophical issues that are not easily discussed. However, it is indeed interesting to consider the theory that the mathematical formula observed in the structure of the pyramid, a deck of cards, and the Universal Calendar is one and the same. One could ask, "Is this all coincidental or providential?" If we assume that the mathematical scheme of the Universal Calendar is somehow inherent in nature, then there is a possibility of pointing to a certain amount of order and design in the universe. The creation accounts in the Hebrew Scriptures, explicitly claim that creation in the beginning should be characterized as order from chaos. The stories "assume a continuing between order and chaos and acknowledge the persistence of evil and the fragility of creation… creation is orderly and deliberate, following a comprehensive plan and resulting in a harmonious and interdependent whole" (Barbour).

(There is another verse I would like to add here. Ephesians 2:20 states, "And are built the foundations of the apostles and the prophets, Jesus Christ himself being the chief cornerstone;" the only building with a "true" chief cornerstone is a pyramid. The very top stone or capstone is the one true chief cornerstone because it is the corner to all four sides of the pyramid.)

"One automatically thinks of the philosophical argument from design," for the existence of God. Admittedly, this is a big jump, but if the mathematical scheme of the Universal Calendar proves to be consistent with the phenomena (such as the pyramids), then Silen may have uncovered a new aspect of the Intelligent Design Argument.

The traditional Design Argument suggests that many things that exist in our universe exhibit order and design. It is claimed that both biological forms and physical conditions favorable for life must be a product of an intelligent designer because it is inconceivable that they could have occurred by chance alone.

The Divine Argument is an argument from analogy. Specifically, it suggests that 1) machines are produced by intelligent design; 2) the universe resembles a machine; therefore, 3) probably the universe was produced by intelligent design.

Of course, many scholars doubt the presupposition that the universe is "like a machine." However, there certainly does seem to be evidence that in many ways there is truth to this theory.

The eighteenth century English theologian, William Paley, one of the major proponents of the Design Argument, compared the universe to a "watch" and claimed that every manifestation of design that exists in a watch also exists in the works of nature. In other words, parts of nature relate to one another in the same way as the parts of a machine relate to one another.

In a watch that is good working order its parts are so connected that when one part moves, the other parts are caused to move as well. Gears are so arranged that the movement of one gear causes another to move. This is a common feature of machines with moving parts, and is also a feature to be found within the universe. Our solar system, for example, is composed of parts, the sun, planets, and their moons, which move, and their moving cause, by gravitational force, other parts to move. Once we understand what a watch is how it works and for what purpose, it would utterly be absurd to suppose that its origin is due to some accident rather than to intelligent design. If we look carefully at many things in nature, plants and animals, for example, we discover that their parts also seem to exhibit an orderly arrangement fitted to a purpose. The implication is that the world of nature arose from intelligent design rather than by accident.

Using a watch as an analogy, here's another: One can take all the parts to make a watch; one of those wind-up watches, and put them in a jar. Shake that jar for the next ten billion years and all the parts will never fall into place to make a working watch. A human or any complex living organism is even more complex than a watch. Now lets say one did manage to eventually get all the parts to fit together to make a working watch out of all that shaking, the next step is to shake the watch in the jar until the watch is wound up. This would be the same as giving the watch life. It all requires an intelligent designer.

In the same manner, the way time has been established in our solar system seems to be by design rather than accident. Paley's "watch" analogy is so important for our purposes because it subtly implies that time itself is a consequence of the order and design of the universe. The Universal Calendar may in fact prove to be a useful (perfect?) tool in measuring time according to a very specific mathematical formula that seems to be inherent at least in our solar system. It seems reasonable to conclude from this that some divine mind designed the universe as a whole, or at least many of it components.

Another personal thought comes to mind on this subject of time. In the Biblical story of Adam and Eve, God told Adam that because he had sinned he would die. Death was God's curse on man. Because Adam knew why, the first question in Adam's mind would probably have been, when? Thus, God's real curse on man is "time." Man has been counting the days ever since. After all, immortality is simply life without "time." It's just a thought.

Since the development of the Darwinian theory of evolution the Design Argument has lost some of its persuasive force. We now posses a fairly well developed naturalistic hypothesis, that explains natural, biological phenomena without the expressed need for an intelligent designer. However, despite the fact that many theologians today implement Darwinian evolution into their theory of how God brings order and design to biological beings, the order and design of the solar system is another issue all together.

The "Big Bang Theory" is today's most promising scientific theory of the origin of the universe. However, even this theory does not take away from the notion of intelligent design. The chance of the universe developing in such away that life is possible in incredibly small, much less than one chance in a billion. As some scientists have persuasively argued, even a small change in the physical constants in the first moments of the Big Bang would have resulted in an uninhabitable universe (Barbour). At the very least, even the Big Bang Theory leaves open the possibility that an intelligent designer had a role in the beginning of the universe, and continues to have a role in the creative energy evident today. Stephen Hawking, for example, writes, "the odds against a universe like ours emerging out of something like the Big Bang is enormous. I think there are clearly religious implications!

British born American theoretical physicist and mathematician famous for his work in quantum field theory, solid-state physics and nuclear engineering,, Freeman John Dyson, born December 15, 1923, gives a number of examples of "numerical accidents that seem to conspire to make the universe habitable."

Is the mathematical scheme of the Universal Calendar evidence of the order of creation, at least in our solar system?

Even if our solar system reflects the only mathematical scheme possible for the existence of life on the planet Earth that does not dismiss the notion of an intelligent designer. As Hawking writes, "what is it that breaths fire into the equations and makes a universe for them to describe?"

In most religions, the notion that God is "perfect" is almost foundational. Certainly, the various religious and philosophical traditions of the world explain this in a variety of ways. However, the presence of a mathematically "perfect" measurer of time (Universal Calendar) may imply divine perfection in a way largely ignored by many theologians today. John Polkinghorne, a physicist and theologian, notes that the key to understanding the physical world is mathematics. Perhaps the key to unlocking the mysteries of divinity is also mathematics. Could mathematical formulas seemingly inherent in the structure of our solar system be the possible linkage between the human mind and the divine intelligent designer?

"Kepler and Galileo, two of the founders of modern science, believed with Plato that God worked according to mathematical models when creating the world. Is it possible that the mathematical equation of the Universal Calendar reveals that the only possible universe (our solar system) is one with constants just like ours? Is it possible that the seven numbers of the Universal Calendar are of divine origin? Is this proof of "divine order" in the universe? These are questions we offer in the hopes of expanding discussion in relation to the old "Argument from Design" or the "Intelligent Design Theory."

In Conclusion: Theological arguments aside, there are very practical reasons for adopting Silen's Universal Calendar. Many of these practicalities are mentioned above. Doesn't it seem odd that human beings seek practicality and convenience in every other realm of life except in the way we structure our time through an antiquated, impractical and inconvenient calendar? In the last century we made immeasurable progress in science and technology.

Our lives benefit from the genius of humanity at every turn, yet we remain bound to a calendar that exists only for the sake of tradition. If we can organize our daily lives with convenient gadget after convenient gadget, why can't we organize our time with a calendar that would simplify existence itself?

Calendar reform leading to a more useable and practical Universal Calendar can come by any one of three ways: 1) Grass-roots support of this calendar; 2) Religious support of the calendar; and 3) Government support of the calendar.

In a changing world, brought together by communication and technological advances, the new millennium would be the perfect time to implement the Universal Calendar. This calendar offers so many practical benefits. Imagine the benefits of all cultures around the world utilizing the same calendar. It would be enormous."

– Dr. Jimmy Watson –

CHAPTER 7
SPEAKING ENGAGEMENTS

I was helping cater an event for a local civic group. This group was made up of almost entirely, senior citizens. I happened to know one of the members and they already knew about the Universal Calendar. He asked me if I would mind giving a lecture to his group, and I accepted his invitation, because I was really curious as to how people would respond in general.

I had never done any public speaking before, but I thought I could handle it. As I got ready to speak, I realized that they were all senior citizens. I figured they were accustomed to their ways and really would reject such a huge change at this time in their lives. I decided not to try and sell them on the idea but to just tell them about the Universal Calendar.

So, I got behind the podium and began to speak.

It began to get extremely hot and I began to sweat profusely; my knees began to shake and I felt weak, however I made it through without fainting. I did come to realize one thing during this lecture, 'I am no public speaker.' I believe that it had to have been the longest thirty minutes of my life. The response after my lecture was totally amazing.

I had some Universal Calendars printed up, and more than 50 percent of the group bought one of the Calendars; the other 45 percent said they wanted to see this calendar adopted before they actually purchased one.

There were only a few people out of the entire group that did not like the Universal Calendar because it would change their birthdays. The response was well over a 95% approval rating. Not in my wildest dreams did I ever expect that high of a positive response. I thought to myself at the time that this was just a fluke response. I decided that I would need a larger audience. But the question now was, how? I did not even like speaking to a relatively small audience.

The county fair was advertising booth space available, so I decided that this would be a perfect opportunity to see how everyone would respond.

I rented a booth and set up my display. I also made a survey sheet that consisted of two columns. On one side I wrote the word, 'agree.' In the other column, 'disagree.' For four days I talked with people about my Calendar, and asked them to take part in a survey. They would sign that they had either agreed or disagreed with this new idea of a reformed calendar.

During those four days the response was totally amazing. Not one single person would sign the disagree side of this survey. These people were from all walks of life. Some of them were doctors and lawyers, but most were just everyday people like me. I have the copies available for anyone who would like to look and prove this true.

The last two days of the fair, I guess I was getting a little excited, and the nervousness seemed to disappear, because I started looking for someone to disagree with the Universal Calendar. I still did not find a single person, not one, that would sign the disagree side.

It seems that this simplistic, logical, and practical little calendar is arguably acceptable with most people.

Could the Universal Calendar actually be a good candidate for adoption? If the Universal Calendar were ever to actually get adopted, it would take a huge majority of support and for people to get involved.

As mentioned earlier, calendar reform is not a new idea, from George Eastman Kodak's attempt, and a few other failed attempts since the early 1920's. I found another article I found a little bit discouraging.

Here is the article titled "U.S. Opposes Action by U.N. on Calendar Reform" from U.S. Department of State Bulletin, April 16, 1955.

The Department of State announced on March 21, 1955, that the U.S. Government had informed the United Nations on that day that does not favor any action by the United Nations to change the present calendar. The United States made its position known in a note transmitted by Henry Cabot Lodge, Jr., U.S. Representative to the United Nations, to the U.N. Secretary General, Dag Hammarskjold, who asked all governments for their views on proposals to revise the existing calendar... The text of the U.S. reply to the Secretary General is as follows:

The Representative of the United States of America to the United Nations present his compliments to the Secretary General of the United Nation and has the honor to refer to the Secretary General's SOA 146/2/01, dated October 7, 1954 concerning World Calendar Reform.

The United States Government does not favor any action by the United Nations to revise the present calendar. This Government cannot in any way promote a change of this nature, which would ultimately affect every inhabitant of this country unless such a reform were favored by a substantial majority of the citizens of the United States acting through their representatives in the Congress of the United States. There is no evidence of such support in the United States for Calendar Reform that is now before the Economic Council. Their opposition is based on religious grounds, since the introduction of a "Blank Day" at the end of each year would disrupt the seven-day sabbatical cycle.

Moreover, this Government holds that it would be inappropriate for the United Nations, which represents many different religious and social beliefs throughout the world, to sponsor any revision of the existing calendar that would conflict with the principles of important religious faiths.

This Government, furthermore, recommends that no further study of the subject should be undertaken. Such a study would require the use of manpower and funds which could more usefully devoted to more vital and urgent tasks. In view of the current studies of the problem being made individually by governments in the course of preparing their views for the Secretary General in 1947, it is felt that any additional study of the subject at this time would serve no useful purpose.

It is exactly this type of statement from this government is exactly what I was afraid of. First, the Universal Calendar does not take away anything from religious views. Second, if the Universal Calendar is to ever be considered for adoption it would take an act of the people writing their representatives and voicing their support.

The Universal Calendar has more meaning than all the other calendars combined throughout history.

The Universal Calendar not only adheres to all religious principles, but may even prove them in mathematics, science, nature, and ecology. The Universal Calendar also applies to economics, entertainment such as music, numerology, tarot, superstitions, and the pyramids.

Yes, the Universal Calendar could affect every inhabitant on the planet, but if it is for the better, why not? This is why public opinion and support of the Universal Calendar is so important.

It is really entirely up to you the reader to decide whether my Calendar is the biggest discovery of the last two thousand years, or the most gigantic coincidence in the history of man. What are the odds of these seven numbers meaning so much.

L.E. Doggett mentions, the legal code of the United States does not specify an official national calendar and stems from an Act of Parliament of the United Kingdom in 1751, which specified the use of the Gregorian calendar in England and its colonies.

He goes on to state that calendar reform would be an extraordinary event, and adoption of a new calendar would depend on the forcefulness with which it introduced and on the willingness of the society to accept it. The United States with its freedom of speech and to choose is the perfect place to introduce a new calendar to the world. Not to mention, the vast resources of media we have in this country.

CHAPTER 8
ASTRONOMY, ARCHAEOLOGY AND SILEN'S UNIVERSAL CALENDAR

Astronomical events, such as the one that took place several years ago – where all of the planets in our solar system (except the Earth) were aligned at the far side of the sun, do not occur very often. Could the combined gravitational pull of all the planets and the Sun, have pulled us just a hair closer to the Sun? And if so—could this have caused what scientists call global warming?

Geological as well as astronomical evidence shows that after this alignment there was an increase in the earth's temperature; an increase in the number and intensity of earthquakes and major storms, such as hurricanes, tornadoes, and tsunamis, and a slight decrease in the distance of the Earth to the Sun. Has it affected the length of a year?

Archeologists dig into the earth to find past civilizations and can tell relatively how long something has been there by the depth in which it was buried. Lets just say for example, archeologists dig 1000 feet down to unearth a civilization which existed more than 1500 years ago. Think about the mega-tons of new weight added to the planet each year. As these layers multiply, one on top of the other, could this also affect the distance of the surface of the Earth to the Sun? Take into consideration of all the extra weight these layers have added throughout the millennia. How does size and weight affect gravity? I'm not a scientist, I don't know, but I do wonder.

One afternoon I decided to conduct a little experiment with the Universal Calendar and the phases of the moon. I took a Gregorian calendar that had the moon phases already on it, and I literally counted the days between each phase of the moon, and placed each phase on the Universal calendar. When I was finished I noticed something rather strange. Each phase of the moon basically had its own specific week. For example, the first week of each month always had a full moon, each second week had a quarter moon, the third week had the new moon, and the forth week, the last quarter moon. It was not always on the same day of the week, but it was the same weeks in every month. I actually figured with the lunar year being eleven days shorter than a solar year that I would run out of lunar phases before I ran out of the year.

I saw in a documentary that in the year 2018, there will be fourteen full moons that year. What really is time anyway? Isn't it just man's concept which developed after observations of the Earth rotating and revolving around the Sun or the moon?

We have broken time down from milliseconds to seconds, minutes, hours, days, weeks, months, years, decades, centuries, millennia, and eons. Now we even have a nanosecond.

If there are indeed, intelligent beings elsewhere in the universe, with its countless stars and planets, it's a pretty safe bet that their concept of time is not the same as ours.

In dealing with, and having come up with this Universal Calendar, many more thoughts about time have come to mind. If one gives a mathematic value to past, present, and future, the only thing to have any value is "now."

For Example the past has already happened and therefore gets a mathematic value of zero, and the future has not happened yet, so it would also get a mathematic value of zero. The only thing that gets any mathematic value is now, and now gets a value of one, because now is the one thing that matters most. One can dig up and even remember the past and contemplate, anticipate, and speculate about future, but the one thing that everyone can absolutely be sure of is "now."

Then there is the ultimate quest of man, since the beginning of time, to find immortality. The answer could lie within our whole concept of time. After all, immortality is really life without the existence of boundaries of time. Time seems to be just a state of mind, like when one is having a lot of fun and paying no attention to time; time just flies by. Watching the clock in anticipation of something, time just drags on and on.

I have unfortunately been in several automobile accidents, and I have also talked with other people who have been in accidents also, and they have also experienced the same thing. Time seemed to be in slow motion. Perhaps time is God's curse on man.

In an immortal setting, time is totally irrelevant.

So if you believe the Universal Calendar is a real discovery that is so simple and logical, and has so much more meaning than all of the past calendars combined, please become an active supporter of the Universal Calendar.

A Positive Scenario for December 21, 2012

There is ONE calendar date that I would like to address; December 21, 2012. If you turn on any documentary channel, whether it is the History Channel or others having to do with 'end-of-the-world' scenarios, one can see all kinds of programs about an up and coming doom, which may take place on that date.

The Mayan Calendar, one of the most astronomically accurate calendars, according to some historians and scientists, stops on December 21, 2012; as does the Chinese I-Ching, and the Hindu calendar. And then there are prophesies. These so-called prophesies from an entire range of persons, such as Nostradamus, Edgar Casey, to Bible figures such as Ezekiel, or John, or Daniel, and Isaiah, as well as others, give bleak pictures of the world's end times.

Let's not forget so-called scientists who jump on board with other doomsday scenarios, ranging from Global warming, Super volcanoes, asteroid impact, Sun flares, and gamma ray bursts, even alien invasions. HUH? Which one will it be? Who knows?

I have more of a positive scenario in mind. December 21, 2012, would make a great starting date toward an improved world. Instituting my Universal Calendar as the world's calendar, may have nothing but positive effect. And Biblically speaking, my Universal Calendar might be a catalyst in opening the heavens to those looking for such things. Such as, the 13th Zodiac. Dr. Michio Kaku, an American Physicist, the Henry Semat Professor of Theoretical Physics at the City College of New York, of City University of New York, the co-founder of the "String field theory" and communicator and popularizer of science, stated that there are actually 13 Zodiacs in the heavens. According to Dr. Kaku, 'Science doesn't know when or why the 13th Zodiac was taken out.'

According to Dr. Shepherd Simpson, "Ophiuchus, the sign of the **Serpent Bearer**, is actually the tenth sign of the Real Solar Zodiac. However, it lies on the ecliptic and is not one of the signs of either the Tropical Zodiac of standard Western Astrology, or the Sidereal Zodiac of the standard Vedic Zodiac astrology. This problem is one of the most contentious in modern astrology.

I know the subject of Ophiuchus causes us lots of grief as astrologers. But it is up there in the heavens. It exists. We just can't ignore it."

Ophiuchus is the Roman 13th Zodiac, and is the Serpent Bearer. However in Greek its name is Asclepius and is the Bearer of the Elixir of Life, and is where the modern symbol of medicine comes from.

Is it also another coincidence that the Egyptian symbol of life, the Christian symbol for life, and the scientific symbol of life, is a CROSS?

In conclusion, many people have asked me, why is instituting my Universal Calendar so important?

All I can say is that it is of course important to me, because it was my discovery – and I stand by my ideas. But why is it important to you? Maybe because we all need hope; maybe it is important because we have all seen too much heartache, and the Calendar may be a symbolic new start to life.

I can say that ONLY YOU may decide if this is important. If truth is as important to you as it is to me then it may well be that important, for I believe the truth about GOD's timing was perfection. Many people who understand the principles behind the calendar have found their own reasons – nearly everyone accepts it once they have seen and heard about it.

Who is in charge of implementing this new calendar – WE, THE PEOPLE!

Unity creates great things, and together we can change our world for the better.

I would like to give special thanks to my personal editor, Dwayne "Deuke" Eukel for all his hard work and effort toward making this book a reality; and to Tracy Lawson, for which this book would not even have been possible. You are both greatly appreciated.